Down to Earth!

Earth Movements

TRACI STECKEL PEDERSEN

PERFECTION LEARNING®

Editorial Director: Susan C. Thies
Editor: Mary L. Bush
Design Director: Randy Messer
Book Design: Michelle Glass
Cover Design: Michael A. Aspengren

A special thanks to the following for his scientific review of the book:
Jason Parkin, Meteorologist, KCCI Weather, Des Moines, Iowa

Image Credits:

Photos.com: cover (bottom), all chapter numbers, some sidebars, pp. 2–3 (background), 4–5 (all), 6 (right), 12–13 (background), 14, 15 (left), 16, 17, 20, 21, 24; Corel: cover (top), title page; MapArt: pp. 7 (right), 8, 10; ClipArt.com: p. 9; South Dakota Tourism: p. 15 (right); Tobi Cunningham: pp. 11, 12, 13; Michelle Glass: pp. 6 (left), 7 (left)

Text © 2006 by Perfection Learning® Corporation.
All rights reserved. No part of this book may be reproduced, stored in a retrieval system, or transmitted in any form or by any means, electronic, mechanical, photocopying, recording, or otherwise, without prior permission of the publisher.
Printed in the United States of America.

For information, contact
Perfection Learning® Corporation
1000 North Second Avenue, P.O. Box 500
Logan, Iowa 51546-0500.
Phone: 1-800-831-4190
Fax: 1-800-543-2745
perfectionlearning.com

1 2 3 4 5 6 PP 10 09 08 07 06 05

Paperback ISBN 0-7891-6573-2
Reinforced Library Binding ISBN 0-7569-4630-1

Table of Contents

1. Get Moving and Change! 4
2. Moving Through Space 6
3. Moving Continents 10
4. Moving Mountains 14
5. Moving Materials 17

Internet Connections
and Related Reading
for Earth Movements 22

Glossary 23

Index 24

chapter 1

Get Moving and Change!

Everything changes. You change. Your friends change. Styles change. Classes change. Cities change. Every day, the world around you continues to change.

The natural world experiences change too. Day turns to night. Seasons come and go. Continents move together and drift apart. Mountains build up and wear down.

Sand piles up into dunes and blows across beaches. Entire landscapes are destroyed and rebuilt. Nothing escapes change.

What causes much of this change? Movement. The Earth moves. The ground moves. Air moves. Water moves. All of this movement brings change to the Earth and to you.

chapter 2 Moving Through Space

Along for the Ride

How much can you do in one second? Snap your fingers? Blink your eyes? Take a step? While you're doing all those things, you're also racing around the Sun at more than 18 miles per second. Actually, the Earth is, but since you're standing on the Earth, you're along for the ride.

How long does it take to complete your journey around the Sun? It takes a little more than 365 days for the Earth to orbit, or go around, the Sun one time.

Leaping Around the Sun

One complete **revolution** around the Sun takes exactly 365 days, 5 hours, 48 minutes, and 45 seconds. Since a calendar year only has 365 days, one extra day is added every four years to even things up (February 29). This fourth year is called a *leap year*.

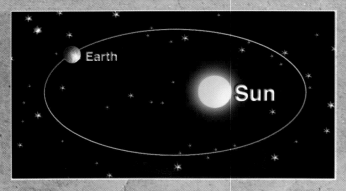

Move Over Summer; Here Comes Winter

The Earth's revolution around the Sun causes the change of seasons. How does this work? The Earth is tilted on an imaginary line called an *axis*. The North and South Poles are at the ends of this axis. The Earth is also divided into Northern and Southern Hemispheres. The two hemispheres are divided by the equator.

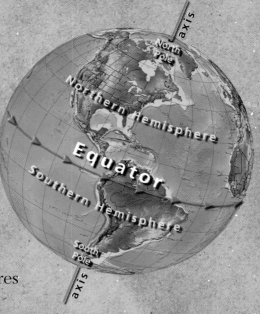

As the Earth revolves, there are times when one hemisphere is tilted toward the Sun while the other is tilted away from it. The hemisphere that is tilted toward the Sun receives more direct sunlight. This half of the planet has summer. At the same time, the opposite hemisphere receives less direct sunlight, so winter occurs there. When neither hemisphere is tilted toward or away from the Sun, spring or fall occurs.

Some places on Earth don't have different seasons. These areas are near the equator. They stay warm year-round because they always receive direct sunlight.

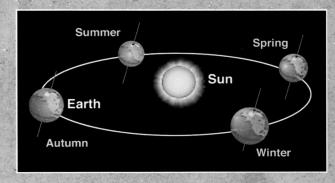

Spinning in Circles

The Earth moves in more ways than one. Besides revolving around the Sun, it also spins on its own axis. The Earth rotates counterclockwise around its axis at 1532 feet per second (at the equator). It makes one full spin, or **rotation**, in a little less than 24 hours.

This spinning causes day and night. As one side of the Earth turns to face the Sun, sunrise occurs there and day begins. The Earth continues turning slowly during the day. As the planet begins to turn away from the Sun, sunset arrives, bringing night. During the nighttime hours, Earth completes its rotation and begins again.

Prove It!

If we're revolving and rotating all the time, why can't we feel it? Our bodies can only sense speed when we're moving faster than the things around us. If you run down the sidewalk, you see the houses pass by. You feel the wind in your face. You can tell that you're moving. As the Earth turns, however, everything is moving at the same speed, so you can't feel it.

Because the movement of the Earth isn't felt, people went thousands of years thinking that the Earth was the center of the universe. People believed that the Earth stood still while the Sun, Moon, planets, and stars passed by. It took centuries and the work of many scientists before it became widely accepted that the Earth was moving.

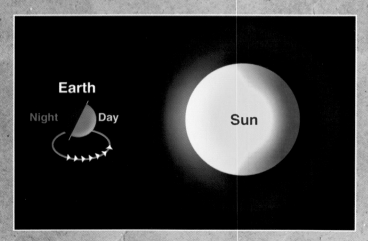

Scientists of Significance

Copernicus

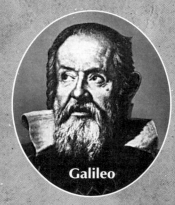
Galileo

Nicolaus Copernicus is considered the Father of Astronomy. In the early 1500s, this Polish **astronomer** claimed that the Earth and other planets revolved around the Sun. He even wrote a book about his ideas. Most people of the time, however, didn't believe his claims, and Copernicus died before he could defend or prove them.

About a century later, an Italian astronomer named Galileo Galilei backed up Copernicus's theories. Galileo was the first scientist to use a telescope to look at bodies in the sky. In 1632, Galileo published his findings. They supported the idea that the Earth revolved around the Sun. At the time, the Catholic Church was very powerful, and Galileo's work went against church teachings. The church forbid Galileo to leave his house for the rest of his life.

Later, the theories of these two important men were proven true. At last their contributions to science were recognized.

chapter 3 Moving Continents

Have you ever noticed how the continents are like giant puzzle pieces? For example, the east coast of South America lines up with the west coast of Africa. About a century ago, a German meteorologist and astronomer named Alfred Wegener noticed this too. He wanted to find out why. After much research, Wegener was convinced that all of the continents had once been one supercontinent that had separated and moved over time.

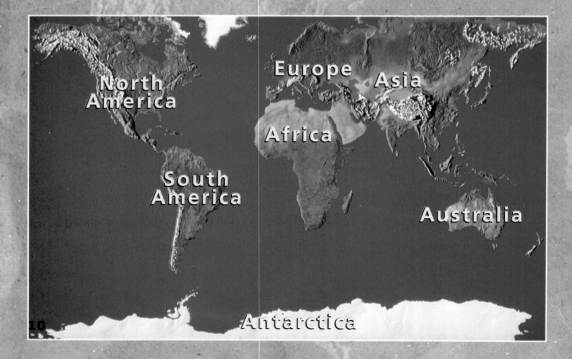

But how? Wegener could not explain how the continents had moved. And without proof, his ideas were not taken seriously. Years later, however, new evidence supported Wegener's theory, which was called *continental drift*. It became widely accepted that the Earth is not one fixed, solid piece. Heat and pressure underground keep the surface moving and changing constantly. This process begins inside the Earth's layers—the inner and outer **cores**, the **mantle**, and the **crust**.

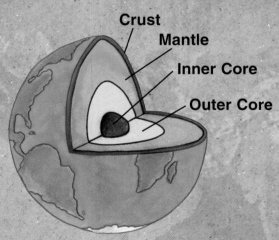

The Core

The center of the Earth is called the *core*. It is divided into the inner and outer cores. The inner core is an extremely hot ball of metal. It is 10,000°F, about the same temperature as the surface of the Sun. The pressure inside the core is also great. Because of this, the liquid metal is squeezed into a solid ball.

The outer core wraps around the inner core. This layer is about 4000–9000°F. Slightly less pressure here allows the metals to flow like a liquid.

The Mantle

Just above the outer core is the mantle. This is the thickest layer of the Earth. It is made of hot, dense rock called **magma**. This rock moves in a slow, circular motion. Hotter areas of rock flow upward, cool, and sink down again.

The Crust

The crust is the Earth's outer layer. It is made of solid rock and **sediment**. Most of the rocks in the crust are igneous rocks. These rocks are formed from cooled magma.

The top of the crust, or surface of the Earth, is mostly covered with sediment and sedimentary rock. Sediment is pieces of rock, **minerals**, shells, and soil. Over time, layers of sediment are pushed together to form sedimentary rock. This rock covers about 75 percent of the Earth's surface.

Metamorphic rocks are igneous and sedimentary rocks that have been changed by heat and/or pressure. For example, when the sedimentary rock limestone is heated and pressed within the Earth, it becomes the metamorphic rock marble.

Moving Through the Cycle

Rocks move through a cycle, changing as they go. Igneous rocks cool at or near the Earth's surface. There these rocks are worn down into sediment. The sediment becomes sedimentary rock, which can change into metamorphic rock. When pushed deep underground, the metamorphic rock melts into igneous rock. Then the rock begins the cycle all over again. In this way, the Earth's surface keeps moving and changing.

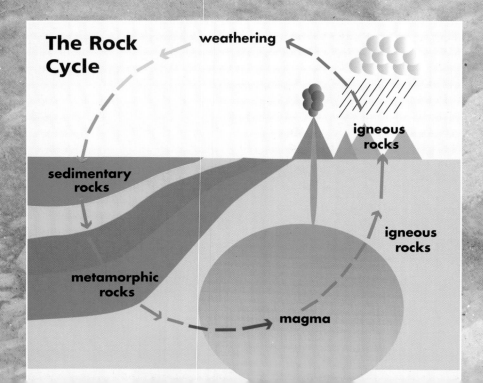

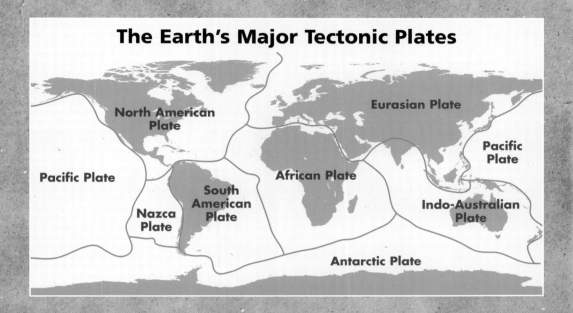

Plate Tectonics

In the 1960s, scientists learned that the Earth's surface is broken up into more than a dozen pieces called **tectonic plates**. Each plate includes the Earth's crust and the top part of the mantle. The plates float and drift on the flowing magma below. They move at about the same speed that your fingernails grow (about 1/2 inch to 4 inches a year).

Plate tectonics was the answer Wegener was looking for! It is now believed that the drifting plates were responsible for moving the continents to their present locations. This continental movement continues today.

chapter 4 Moving Mountains

Mountains cover one-fifth of the Earth's land. They are home to one-tenth of the world's people. How did these beautiful, impressive features come to be? Through the Earth's movement, of course.

Plate movement and underground forces make powerful changes in the Earth's landscape. Depending on the kind of movement, five different types of mountains are formed.

Fold Mountains

When two tectonic plates push against each other and neither gives, the land buckles up and folds over into a mountain range. These are called *fold mountains*. The Himalayan Mountains in Asia and the Appalachian Mountains in the eastern United States are fold mountains.

Himalayas

Fault-Block Mountains

When moving plates stretch and crack rocks in the crust, they create a **fault**. Large sections of solid rock can push up through this opening. These rocks become fault-block mountains. The San Andreas Fault in California is responsible for the fault-block mountains known as the Sierra Nevada.

Sierra Nevada

Dome Mountains

Sometimes magma pushes against the Earth's crust but doesn't actually break through. Instead, the magma hardens and pushes upward on the land. Eventually, the ground above becomes a dome-shaped mountain. The Black Hills of South Dakota are dome mountains featuring **caves**, cliffs, canyons, and waterfalls.

Black Hills

Plateau Mountains

A plateau is a flat area of land that is raised above surrounding land. Normally plateaus are stacks of sedimentary rock. When rivers or glaciers wear down the rock, plateau mountains are formed. The Catskill Mountains in New York are plateau mountains carved by glaciers millions of years ago. Today, rivers and streams running through the Catskills continue to **erode** the mountains.

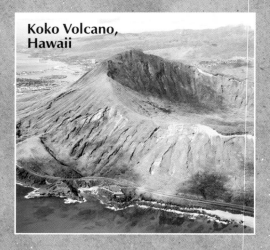

Koko Volcano, Hawaii

Volcanic Mountains

Plates that move away from each other form long, deep cracks in the crust called *rifts*. Magma wells up through these rifts. Eventually it explodes out of the ground as **lava**. When the lava reaches the Earth's surface, it cools quickly and hardens into rock. As lava builds up on either side of a rift, volcanic mountains are born. Each time magma erupts from the ground, a volcanic mountain grows and changes.

Most volcanic mountains begin in the ocean. The Mid-Atlantic Ridge is in the Atlantic Ocean. This ridge is more than 10,000 miles long. It is the longest mountain range in the world.

Technology Link

Scientists continue to work to improve their methods for monitoring volcanoes. Scientists at the Hawaiian Volcano Observatory are among the best. Their advanced instruments and measuring techniques are now used to study **active** volcanoes all over the world.

Visit the observatory at **http://hvo.wr.usgs.gov/**.

Chapter 5: Moving Materials

Have you ever left your bike outside in the sun or rain for a long period of time? After a while, did the bike begin to look different? Maybe its color became dull or the metal parts got rusty.

A bike left outside is like anything else in nature. When exposed to sunlight, heat, rain, wind, and ice, things move and change. These changes are caused by **weathering** and **erosion**.

Weathering: Falling to Pieces

Weathering is the wearing away and breaking down of rock and soil. There are two main types of weathering—physical (or mechanical) weathering and chemical weathering.

Physical Weathering

Physical weathering breaks down rock without changing what it's made of. Small pieces of rock get chipped, cracked, peeled off, or worn down. But no changes occur in the minerals that make up the rock.

Many rocks allow water to seep through them. They have cracks, joints, or tiny holes that water can flow through. In colder temperatures, water stuck inside rocks freezes into ice. Because ice takes up more space than water, it pushes against the tight spaces. The cracks grow larger to make room for the ice. When the ice melts, the bigger cracks and holes allow more water to enter, freeze again, and push on the rock even more. This process is called the *freeze-thaw cycle*. Over time, the spaces get so big that the rock splits apart.

The minerals in rocks expand (get bigger) when they heat up and contract (get smaller) when they cool off. In places where temperatures get very hot during the day and very cool at night, rocks are constantly expanding and contracting. Eventually, this process weakens the rocks, and outer layers peel or break off.

Wind also plays a part in the weathering process. Strong gusts of wind can pick up small grains of sediment and blast them against rocks. Over time, this wears down rocks and carves landforms.

Water can also weather rock and sand. Strong waves beating on rocky cliffs or sandy beaches smash rocks and wash away surfaces. Large masses of frozen water called *glaciers* scrape and break rocks in their paths.

Plants and animals also cause weathering. Tree and plant roots work their way into cracks in rocks. As the roots grow bigger, they can split the rocks apart. Animals that burrow into the ground also break up rocks in their way.

Chemical Weathering

Chemical weathering causes the minerals inside a rock to change. Water is the main cause of this type of weathering. Rain absorbs carbon dioxide gas from the air. This creates a weak acid that slowly eats away at rock.

In areas where air pollution is thick, acid rain may form. This rain absorbs harmful gases in the air and becomes even more destructive. Acid rain speeds up the weathering of rocks, statues, and buildings.

Inquire and Investigate: Chemical Weathering

Question: How is weathering by chemicals different than weathering by water?

Answer the question:
I think chemicals weather rocks _____.

Form a hypothesis: Chemicals weather rocks (faster/slower) than water.

Test the hypothesis:

Materials
- limestone or marble chips
- 2 small glass jars
- water
- vinegar
- 2 aluminum pie pans

Procedure
- Place an equal number of limestone or marble chips in each of the jars.
- Cover the rocks in one jar with water.
- Cover the rocks in the other jar with vinegar (an acid).
- Use just enough liquid to cover the rocks.
- Let the jars sit overnight.
- The next day, pour the liquid from each jar into the pie pans.
- Discard the rock chips.
- Allow the liquid in the pans to evaporate.
- Observe the amount of solid material left in the pans.

Observations: The vinegar pan will have more solid material in it than the water pan.

Conclusions: Chemicals weather rocks faster than water. There's more rock material in the vinegar pan because the vinegar (acid) weathered the rock chips more than the water did. This shows that the weathering process occurs more quickly with chemicals than with water.

Erosion: Getting Carried Away

Weathering breaks down materials. Erosion moves these materials from one place to another. The main forces of erosion are wind, water, and ice.

Wind

Wind is a powerful force that both wears down rock and carries the particles away. The blowing particles then help the wind weather new rock in its path. Gusting winds also pick up sand grains and deposit them on beaches or in dunes.

Water

Rivers and streams pull rocks and sediment with them. Heavy sediment is deposited along the way. Lighter sediment may be carried all the way to the ocean.

Fast-moving rivers carve valleys and canyons out of rock. The Grand Canyon in Arizona was carved by the roaring Colorado River.

Rivers and streams under the Earth's surface have the same weathering effects as above ground. Limestone rock is especially vulnerable to **groundwater**. Over time, water reshapes the limestone rock, carving out passageways and caves.

Carlsbad Caverns, New Mexico

Sea cave on the coast of Turkey

Once sediment reaches the ocean, some of it settles on the ocean floor. The rest is carried by water **currents**. Every day, the **tides** wash sediment ashore and drag it back into the water. This daily action both wears away and adds to the coastline, changing its shape.

Rocky cliffs along the ocean are also eroded by crashing waves. Sometimes the erosion is so great that sea caves are formed.

Ice

Glaciers are slow-moving masses of ice that are pulled downhill by gravity. Most glaciers move just several feet in a year. Occasionally, however, a glacier will surge more than 100 feet in a day!

As a glacier slides, it picks up pieces of rock, sand, and soil. These materials then travel with the huge chunk of ice. Some is deposited along the glacier's path. The rest is carried away in the water that flows when the glacier melts.

Caves can also form when running water moves beneath glaciers. These tunnels of water carve out ice caves.

On the Move

Earth is always on the move. It is turning, pushing, growing, and shrinking. So the next time you watch a sunset or feel the sand between your toes, remember how the Earth's movements change your world.

Internet Connections and Related Reading for Earth Movements

http://kids.msfc.nasa.gov/Earth/
Learn more about the Earth, plate tectonics, and the Sun at this NASA site for students.

http://www.allaboutnature.com/subjects/astronomy/planets/earth/seasons.shtml
This simple site introduces the seasons, the tilt of the Earth's axis, and how fast the planet moves.

http://science.howstuffworks.com/volcano.htm
Moving plates, erupting magma, hot rock—find out how all of these elements work together to create volcanoes.

http://www.minsocam.org/MSA/K12/K_12.html
Review the rock cycle, the three types of rocks, and how plate tectonics "recycles" the rocks on Earth.

* * * * * * * * * * * *

How Mountains Are Made by Kathleen Weidner Zoehfeld. Describes plate tectonic theory and how the forces of nature shape our world. HarperCollins, 1995. [RL 2 IL K–4] (4799601 PB 4799602 CC)

Sunshine Makes the Seasons by Franklyn M. Branley. Describes how sunshine and the tilt of the Earth's axis are responsible for the changing seasons. HarperCollins, 1985. [RL 2 IL K–4] (8746401 PB 8746402 CC)

Volcanoes: Mountains That Blow Their Tops by Nicholas Nirgiotis. Describes the formation and activities of volcanoes and identifies some notable eruptions. Grosset Dunlap, 1996. [RL 2.8 IL K–3] (6914601 PB 6914602 CC)

What Makes Day and Night by Franklyn M. Branley. An excellent science book for beginning readers that explains the revolution of the Earth. HarperCollins, 1986. [RL 2 IL K–3] (9838301 PB 9838302 CC)

- RL = Reading Level
- IL = Interest Level

Perfection Learning's catalog numbers are included for your ordering convenience. PB indicates paperback. CC indicates Cover Craft.

Glossary

active (AK tiv) still erupts occasionally

astronomer (uh STRAHN uh mer) person who studies the skies (Sun, Moon, planets, stars, etc.)

cave (kayv) large, naturally hollowed-out place in the ground, rock, or ice

core (kor) center of the Earth divided into inner and outer layers

crust (kruhst) outer layer of the Earth

current (KER ent) steady flow of water in a particular direction

erode (uh ROHD) to wear down and carry away pieces of rock, sand, or soil

erosion (uh ROH zhuhn) movement of rock, sand, and soil by wind, water, or ice

fault (fawlt) break in the Earth's crust due to stress

groundwater (GROWND waw ter) water that soaks into the soil and collects underground

lava (LAH vah) liquid rock above the Earth's surface

magma (MAG mah) liquid rock within the Earth

mantle (MAN tuhl) layer of the Earth between the crust and core (see separate entries for *crust* and *core*)

mineral (MIN er uhl) nonliving substance, such as copper, gold, or silver, that makes up rock

plate tectonics (playt tek TAHN iks) study of the movement of the Earth's plates

revolution (rev uh LOO shuhn) movement of one object around another (Earth around the Sun)

rotation (roh TAY shuhn) movement of an object around a fixed point (Earth around its axis)

sediment (SED uh ment) small pieces of rocks, minerals, and soil (see separate entry for *mineral*)

tectonic plate (tek TAHN ik playt) piece of the Earth's crust (see separate entry for *crust*)

tide (teyed) rise and fall of the ocean due to the Moon's gravity that occurs every 12 hours

weathering (WETH er ing) process of breaking down rocks into smaller pieces

Index

continental drift, 10–11

Copernicus, Nicolaus, 9

Earth's revolution, 6, 7, 8, 9

Earth's rotation, 8

erosion, 20–21
    ice, 21
    water, 20–21
    wind, 20

Galilei, Galileo, 9

layers of the Earth, 11–12
    core, 11
    crust, 11–12
    mantle, 11

mountains, 14–16
    dome, 15
    fault-block, 15
    fold, 14
    plateau, 15
    volcanic, 16

plate tectonics, 13

rock cycle, 12

weathering, 17–19
    chemical, 18, 19
    physical, 17–18, 19

Wegener, Alfred, 10, 11, 13